LIFE SCIENCE IN DEPTH

PEMBROKESHIRE COLLEGE
LEARNING CENTRE
HAVERFORDWEST
PEMBROKESHIRE SA61 1SZ
01437 765247
www.pembrokeshire.ac.uk

an or b... ...te stamped below.

B~~~~~~~~~~~~S

AND

HEALTH

Ann Fullick

D0308999

PO31298/PO31321

www.heinemann.co.uk/library
Visit our website to find out more information about Heinemann Library books.

To order:

☎ Phone 44 (0) 1865 888066
📄 Send a fax to 44 (0) 1865 314091
💻 Visit the Heinemann bookshop at www.heinemann.co.uk/library to browse our catalogue and order online.

First published in Great Britain by
Heinemann Library, Halley Court, Jordan Hill,
Oxford OX2 8EJ, part of Harcourt Education.

Heinemann is a registered trademark of
Harcourt Education Ltd.

© 2006 Harcourt Education Ltd.
First published in paperback in 2007.

The moral right of the proprietor has been
asserted.

All rights reserved. No part of this publication
may be reproduced, stored in a retrieval system,
or transmitted in any form or by any means,
electronic, mechanical, photocopying, recording,
or otherwise, without either the prior written
permission of the publishers or a licence
permitting restricted copying in the United
Kingdom issued by the Copyright Licensing
Agency Ltd, 90 Tottenham Court Road, London
W1T 4LP (www.cla.co.uk).

Editorial: Sarah Shannon and Dave Harris
Design: Richard Parker and Q2A Solutions
Illustrations: Q2A Solutions
Picture Research: Natalie Gray
Production: Chloe Bloom

Originated by Modern Age Repro
Printed and bound in China by South China
Printing Company

10 digit ISBN: 0 431 10897 8 (hardback)
13 digit ISBN: 978 0 431 10897 1

10 digit ISBN: 0 431 10904 4 (paperback)
13 digit ISBN: 978 0 431 10904 6

11 10 09 08 07
10 9 8 7 6 5 4 3 2 1

British Library Cataloguing in Publication Data
Fullick, Ann, 1956-
  Body systems and health.
  - (Life science in depth)
  571
A full catalogue record for this book is available
from the British Library.

Acknowledgements
The publishers would like to thank the following
for permission to reproduce photographs:
Alamy p. **45** (Popperfoto); Anthony Blake
p. **14** (Sian Irvine); Corbis pp. **11** (Bettmann),
22 (Elizabeth Hathon), **38** (Jennie Woodcock),
54 (Martin Harvey/Gallo Images); Digital Vision
p. **12**; Empics/PA p. **33** (Andrew Parsons);
Getty Images pp. **4** (Bob Thomas), **53** (Bruce
Ayers), **37**, **56** (PhotoDisc); Mary Evans p. **41**;
PhotoLibrary.com p. **44**; Science Photo Library
pp. **52** (Biology Media), **48** (Christian Jegou),
5 (CNRI), **20** (Du Cane Medical Imaging ltd),
28, **46**, **50** (Eye of Science), **40** (George
Bernard), **42** (J.C. Revy), **36** (John Cole),
10 (K.Ward/Biografx), **1**, **30** (Neil Bromhall),
59 (Simon Fraser).

Cover photograph of a scan of the veins and
arteries around the heart, reproduced with
permission of Science Photo Library/Zephyr.

Our thanks to Emma Leatherbarrow for her
assistance in the preparation of this book.

Every effort has been made to contact copyright
holders of any material reproduced in this book.
Any omissions will be rectified in subsequent
printings if notice is given to the publishers.
The paper used to print this book comes from
sustainable resources.

Contents

PEMBROKESHIRE COLLEGE LIBRARY

Words printed in the text in bold, **like this**, are explained in the Glossary.

Systems of life

The human body comes in all shapes and sizes. Our bodies start working before we are born and keep going until we die, sometimes as much as one hundred years later. Yet mostly we take them completely for granted. If we look a little closer, there are lots of amazing things to discover about what goes on inside our skin.

BODY ORGANIZATION

The human body is made up of millions of tiny cells. These cells are not all the same; they are **specialized** to carry out different jobs within the body. But the body is not just a solid mass of cells. The cells are arranged into **tissues**, and then the different tissues are arranged into **organs**. Some of the organs that make up your body, such as the heart and the kidneys, are well known. Others, like the pancreas and the spleen, might not be familiar to you, but they are needed just as much for your body to work well.

The organs are organized into systems, where several of them work together to carry out an important function.

Everyone looks different. But it doesn't matter what we look like, our bodies all work in the same way.

The **digestive system** breaks down the food we eat, the **circulatory system** carries oxygen, food and waste around our bodies and the reproductive system makes new human beings. There are many other organ systems too, all working together to keep us fit, healthy – and alive!

KEEPING HEALTHY

Most babies are born with healthy, working bodies, but not everyone is so lucky. Sometimes things go wrong as a baby develops in its mother's body, and sometimes the **genetic information** passed from the parents to the child is faulty so that one or more of the organ systems does not develop properly.

Even if we start off with a healthy body, the choices we make about the way we live can affect and damage our bodies. This can make us ill and even threaten our lives. Understanding how our organ systems work and looking after our bodies can help to keep us active and healthy, living life to the full.

In the living body, all the organ systems are packed together and held in place by muscles and skin. This is a scan of the abdominal organs.

Did you know..?

It can be surprising how much the different organs of the body weigh. The brain and liver weigh about 1.4 kilograms (3 pounds) each while the spleen weighs about 0.2 grams (0.44 pounds). But the heaviest organ is the skin. It weighs about 16 percent of the total body weight, so the skin of an average 70 kilogram (154 pound) man weighs almost 13 kilograms (29 pounds)!

Body fuel

All living things need energy to grow, move and reproduce. Plants make their own food using energy from the Sun, carbon dioxide, and water. But animals can't do this – they need to eat plants or other animals to get their food.

Humans don't just need food to provide energy. We also need food to give us the building blocks to grow and repair the bits that go wrong or wear out.

FOOD, GLORIOUS FOOD...

The food we eat varies enormously depending on where we live. In countries like the United States and the United Kingdom, people eat a wide range of foods including meat, bread, fruit and vegetables, cheese, eggs, and milk, and most people have plenty to eat. In many other countries there is less food and less variety. Diets are often based around a cereal such as rice or maize with a few vegetables and a little meat. Whatever we eat, our bodies need certain things to stay healthy. A balanced diet contains:

Carbohydrates – energy-rich foods the body can use easily (for example, bread and rice)

Proteins – the building blocks the body needs to grow and repair itself (for example, meat and fish)

Fats – high-energy food needed for cell **membranes**, the nervous system, and to build an energy store; too much fat can cause people to become overweight (for example, butter and cheese)

Vitamins – compounds such as vitamin A and C, needed in very small amounts to keep our bodies healthy (found in fruit, vegetables, meat, and dairy products)

Minerals – such as sodium, calcium, and iron, needed in very small amounts to keep our bodies healthy (found in meat, dairy products, fruit, nuts, and vegetables)

Fibre – not **digested** but keeps the gut moving (comes from a substance called cellulose in cell walls of plants)

Water – no cell can work without water – all living organisms are made up of about 80 percent water. We get it from drinking and from our food.

CLASSIC EXPERIMENT vital vitamins

In the early 20th century, Sir Frederick Gowland Hopkins carried out some classic experiments at Cambridge University. He took two groups of young rats and fed both of them special food made of protein, carbohydrate, and fat, along with mineral salts and water. One group was given an added extra. They got 3 cm^3 of milk every day for the first eighteen days of the experiment. Then their milk supplement was stopped, and the extra milk was given to the other group until the end of the experiment. As this graph shows us, the vitamins in the milk were vital for the young rats to grow. Whichever group was getting the milk supplement grew really well, but without the milk, they hardly gained weight at all.

Without milk, Gowland Hopkins' rats just didn't grow. Giving them a tiny quantity of milk made all the difference.

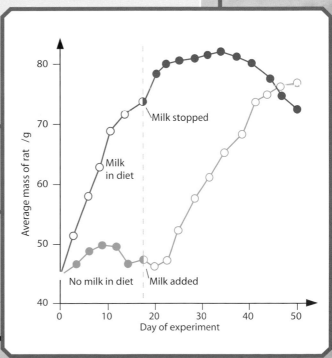

Average mass of rat /g

Milk stopped

Milk in diet

No milk in diet   Milk added

Day of experiment

THE JOB OF THE DIGESTIVE SYSTEM

We eat our food in bite-sized chunks, and at this stage it is made up mostly of very big **molecules** of carbohydrate, protein, and fat. However, our bodies need small molecules that can be carried around in our blood and move into the individual cells where they are needed. These small molecules can then be broken down to release energy, or built up into the big molecules we need to make our cells. We have a complicated set of organs to take the food we eat and break it down into the small, useful molecules we need. It is called the digestive system (or gut).

THE GUT STARTS HERE!

Your digestive system is a long, hollow muscular tube that runs right through your body from your mouth to your **anus**. It is about 9 metres (30 feet) long from beginning to end. Food is pushed along the tube by muscle contractions known as **peristalsis**, and it is digested (broken down) as it goes. You don't actually take food into your body until the digested molecules move out of the hollow tube of your gut and are absorbed into your blood.

Your digestive system breaks down food in two different ways.

- It breaks food up physically into smaller pieces which are easier to swallow. Lots of small pieces also have a very big surface area, which makes it easier for the digestive juices to reach all of the food and work on it.
- It breaks food up chemically from big molecules into smaller molecules that your body can use. This process is called digestion.

THE DIGESTION MACHINE

The food we eat takes around 24 hours to pass through our complicated digestion machine. The toast you eat at breakfast will have been broken down and absorbed into your bloodstream by the afternoon, but any indigestible waste would probably sit in your gut waiting to be removed the next morning.

A drawing of the human digestive system looks quite simple, but in the living body there are metres of muscular tubes all moving about in a very small space.

The mouth, where the teeth and the saliva start the process of digestion.

The oesophagus is a muscular tube about 25 centimetres (10 inches) long. Food is squeezed along by peristalsis, so we can swallow even when we are upside down!

The stomach is a strong muscular bag that contains hydrochloric acid and protein-digesting enzymes. It squeezes food into a thick paste called chime.

The liver makes alkali, which neutralizes the stomach acid, and it also makes bile, which helps to digest fats.

The gall bladder stores the bile made by the liver and releases it into the gut along the bile duct.

The pancreas is tucked under the stomach, and makes lots of digestive enzymes which go into the small intestine.

The small intestine is the longest part of the gut – about 7 metres (23 feet) long. It contains enzymes that break down carbohydrates, proteins and fats. This is also where the small molecules produced by digestion are finally taken into the bloodstream, along with minerals and vitamins.

The large intestine contains all the waste, undigested food, and lots of water. The waste moves through the large intestine while water is taken back into the body.

The anus is the ring of muscle that closes the gut at the bottom end. Faeces (solid waste) is passed out of the anus.

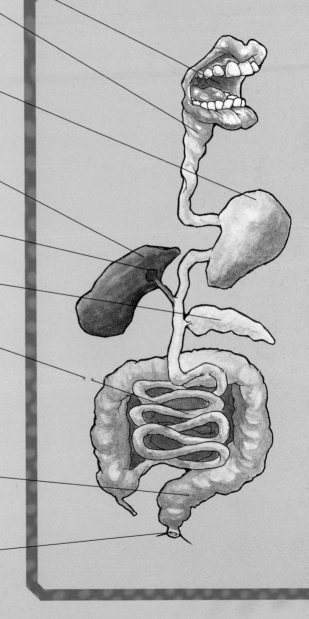

THE BREAKDOWN STORY

The food you take in is made up of very big, complicated molecules, which your body needs to build cells and get energy. The problem is that these large food molecules are just too big to pass out of your gut and into your blood stream. Left alone in your gut, it would take days for food to be broken down by the acid in your stomach and the action of **bacteria** and **fungi** in your gut. In the meantime, you would starve to death as you waited for the energy to be released.

To stop this from happening, your digestive system makes a range of special molecules known as **enzymes**. These enzymes speed up the whole breakdown process. In just a few hours, food that we eat is broken down into the molecules we need, and arrives in our bloodstream. These molecules include **glucose**, **amino acids**, **fatty acids**, and **glycerol**.

WHAT IS AN ENZYME?

Most enzymes are proteins. They are complicated molecules known as biological catalysts, special molecules which speed up the chemical reactions in cells. They work by making it easier for the chemical reactions of digestion to take place, so everything gets broken down faster.

Enzyme molecules are very complex proteins. They have special active sites, where particular molecules such as carbohydrates can fit. The molecules are bound to the enzyme at the active site, so the enzyme can do its job.

Different types of enzymes speed up the digestion of different food substances. For example:

- amylase, found in saliva and in the small intestine, breaks down carbohydrates into simple sugars such as glucose
- pepsin, found in the stomach, and peptidase in the small intestine, break down proteins into amino acids
- lipase, found in the small intestine, breaks down fats into fatty acids and glycerol.

Being proteins, enzymes are very sensitive to changes in the temperature and acidity of their surroundings. So the conditions in the different parts of the digestive system all help to ensure the enzymes work properly. The enzymes need the acid in the stomach, the alkali in the small intestine, and the constant body temperature of around 37 °C (99 °F).

SCIENCE PIONEERS enzymes are proteins!

In 1926, the American, James Sumner, produced the first pure enzyme crystals. He extracted them from jack beans. As the crystals were protein, he suggested that all enzymes must be protein, but nobody believed him.

Between 1930 and 1936, John Howard Northrop, another American chemist, extracted pure versions of the protein-digesting enzymes pepsin, trypsin, and chymotrypsin from the gut, and showed that enzymes were definitely proteins. He received a Nobel prize for his work.

This is a picture of some of the Nobel Prize winners in 1946. Dr John Howard Northrop is standing third from the left.

YOU ARE WHAT YOU EAT!

As famed American novelist Mark Twain said, "Part of the secret of success in life is to eat what you like and let the food fight it out inside". However, he also said "The only way to keep your health is to eat what you don't want, drink what you don't like, and do what you'd rather not." Most of us would like to follow Twain's first piece of advice and ignore the second. Unfortunately, more and more research shows that the food we eat does have a major effect on our health.

Obesity is a growing problem – and it isn't just adults who are becoming hugely overweight. In many countries obese teenagers are becoming worryingly common.

TOO MUCH FOOD...

Everyone needs to eat a certain amount every day to give them the energy to stay alive and to provide the energy for everything else we do, so very active people need to eat more than less energetic friends. But if we take in more energy than we use up, our bodies store the extra energy as fat. One of the biggest problems in countries such as the United States and the United Kingdom is that too many people are eating too much food, especially fatty and sugary food. These foods taste good and are cheap and easy to buy, so people eat more of them than they used to. As a result, people are steadily getting fatter and fatter. This can lead to all sorts of health problems.

... AND TOO LITTLE FOOD

In many countries the problem is not too much food, but too little. In areas of Africa and Asia, the rains are failing year after year. This means the crops fail too and everyone goes hungry. Swarms of **locusts** can destroy every plant for hundreds of miles, while civil wars and violence mean people can no longer tend their fields safely. All of these things mean famine and starvation for the local people. When the body is deprived of food for a long time, not only does the person become very thin, they can also suffer from a range of illnesses due to lack of protein, vitamins, and minerals in their diet. Sores break out, their bones become weakened, and almost every system of the body is affected by this **malnutrition**.

Did you know..?

The amount of energy contained in our food is measured in kilojoules or calories. The average teenage boy needs about 11,510 kJ (2,750 calories) a day while teenage girls need around 8,830 kJ (2,110 calories). Energy used to be measured in kilocalories (also called calories). Scientists always use kilojoules, but lots of ordinary people still think of the energy in food in terms of calories. Most food packets show you the energy measured in both.

EAT WELL FOR HEALTH

Eating sensibly and well has an effect on your whole body. Examining some of the health problems that are linked to the way people eat shows just how important that can be.

CANCER AND DIET

Scientists are still a long way from understanding cancer. Many different factors determine whether someone will develop cancer, including genetic information inherited from their parents. However, evidence is growing that what you eat makes a difference. Eating five portions of fruit and vegetables each day seems to lower the risk of developing many different types of cancer (as well as heart disease). Eating plenty of cooked tomatoes seems to lower the risk of prostate cancer, while the diet of the Japanese (which is very high in soya bean products and green tea) seems to protect them against breast cancer.

PROBLEMS CAUSED BY FOOD

Not eating enough, or eating the wrong foods, can also cause big problems. Your diet directly affects your teeth. Sugary foods and drinks provide food for the bacteria which cause tooth decay. Overeating and obesity are linked to a whole range of health problems such as diabetes (when you can't control the sugar levels in your blood), heart disease (see page 46), high blood pressure, and breathing problems such as asthma.

A healthy diet may be the best defence against a wide range of diseases. Making the right choices in the food we eat is very important.

A lack of minerals or vitamins in the body can cause a range of deficiency diseases, some of which are shown in the table below.

Vitamin/ mineral	Common source in food	Deficiency diseases
Iron	Red meat, eggs, apricots, cocoa	Anaemia – blood doesn't carry oxygen properly causing tiredness and weakness
Vitamin B1	Yeast, cereals	Beri-beri – muscle wasting, stomach upsets, circulation failure, paralysis
Vitamin C	Citrus fruits	Scurvy
Vitamin D	Fish liver oil	Rickets – bones are soft and don't grow properly

RECENT DEVELOPMENT the diet for long life?

The Japanese island of Okinawa has a very high number of healthy people over 100 years old. The common diseases of old age are very rare on the island. Scientists have been studying the islanders closely, and the secret of their long lives seems to be their traditional diet and lifestyle. They eat relatively little, and their food is high in soya, fruit, and vegetables, and low in fat. They are physically active and are generally happy in their life. If they move away from the island and the traditional ways of eating and living, they lose the health benefits – more evidence that in Okinawa health is linked to lifestyle.

Support, protect, move

Standing and moving around are second nature to most people. We don't think about it, in the same way that we don't have to think about what shape our body is as we go to school or run down the street. Yet without our skeletons and muscles, we would be shapeless blobs – and we would not be able to move!

THE SKELETON

The skeleton itself is made up of lots of different bones. A bone is a combination of living bone cells set into a matrix (mesh) of special protein fibres which are hardened by calcium salts. The calcium salts make the bone very strong, while the cells and proteins stop it from being brittle and breaking too easily. A solid skeleton would be so heavy we wouldn't be able to move around, so your bones are hollow. There are two types of bone. Compact bone is very heavy and dense and it is found where extra strength is needed. The rest is spongy bone, which has a lighter, open structure.

CLASSIC EXPERIMENT

If a bone is placed in hydrochloric acid, it will fizz as the calcium salts react with the acid. Once the reaction stops, the bone can be removed from the acid and washed. It is no longer rigid. Without the calcium salts it becomes a relatively soft, floppy material that cannot support the weight of a body. But if the bone is treated to remove the protein while the calcium salts remain, the bone becomes so brittle that it crumbles at a touch. This experiment shows that the calcium salts and protein are both very important for a bone to be strong but not too brittle.

THE ROLE OF THE SKELETON

The skeleton has three important functions in your body:

- it supports your body against gravity and allows you to stand up and keep your body shape.
- it protects the delicate internal organs against damage – for example, the ribs protect your heart and lungs.
- the joints of the skeleton allow you to move – different types of joints make different types of movement possible.

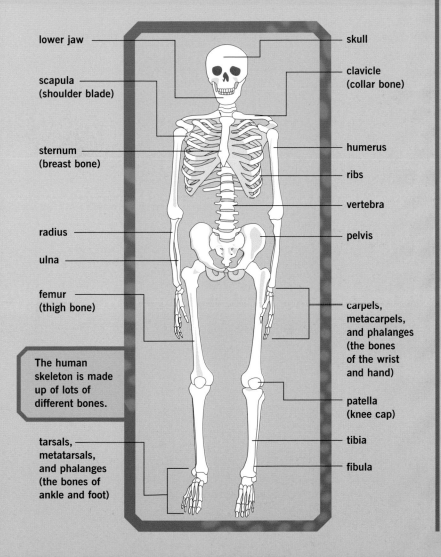

lower jaw

skull

scapula (shoulder blade)

clavicle (collar bone)

sternum (breast bone)

humerus

ribs

vertebra

radius

pelvis

ulna

femur (thigh bone)

carpels, metacarpels, and phalanges (the bones of the wrist and hand)

The human skeleton is made up of lots of different bones.

patella (knee cap)

tarsals, metatarsals, and phalanges (the bones of ankle and foot)

tibia

fibula

MOVING ABOUT

When it comes to moving about, the joints are very important. A rigid skeleton would support and protect us but it wouldn't be much use if we couldn't move. And moveable joints are useless if they can't lift the weight of the bones. This is where the other tissues of the skeletal system play their part. Muscles, cartilage, ligaments, and **tendons** all help us to move around.

THE STRUCTURE OF A JOINT

A joint is where two bones meet and move against each other. In the joint, the surfaces of the bones are covered with a layer of smooth, rubbery tissue called cartilage. This stops the bones from grinding together and wearing away. The cartilage layer cushions the joint from the jarring when you run and jump as well.

In joints like the hip, shoulder, and knee, which move a lot, cartilage alone isn't enough to protect the joint from wear and tear. A special membrane in these joints produces an oily liquid called synovial fluid which lubricates the joint and keeps it moving smoothly.

To make sure that the bones stay in the right position as the joint moves, they are held together by strong fibres called ligaments.

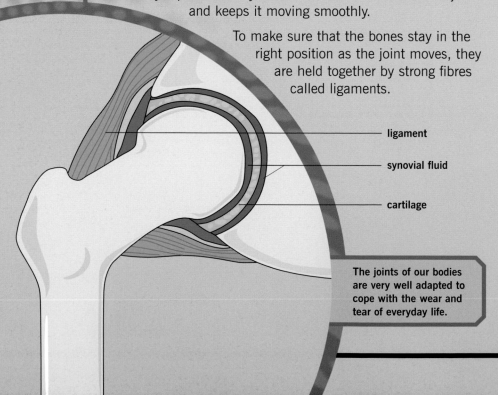

ligament

synovial fluid

cartilage

The joints of our bodies are very well adapted to cope with the wear and tear of everyday life.

ON THE MOVE

The actual movement of the bones is brought about by muscles. Muscles are made up of long cells called muscle fibres. These are made of protein and held together in bundles. Muscle fibres can contract (shorten) and then relax again. When muscles are working hard, they need lots of oxygen and glucose, so they have a good supply of blood.

Muscles are attached to the bones by tendons, which don't stretch at all. This means that when the muscles work, all the effort goes into moving the bones.

Muscles can only pull on a bone to move it. They cannot push it back into its original position. Another muscle is needed to pull the bone back. So muscles usually occur in pairs, known as antagonistic pairs because they work in the opposite direction to each other. If one muscle pulls a bone up, another muscle pulls it down again.

A well-known example of an antagonistic pair is the biceps and triceps muscles in the upper arm. Lift your arm up and watch the biceps bulge. Then when you lower your arm, the biceps relax while the triceps under your arm tense up to control the downward movement.

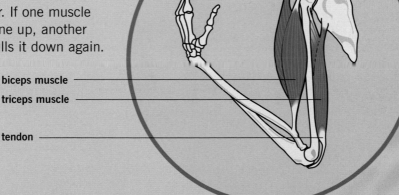

biceps muscle

triceps muscle

tendon

Did you know..?

In a fit adult, muscles can make up to 40 percent of the body weight!

MAKING AND BREAKING

The skeleton and the muscles both respond to exercise. When you walk and run, it builds up the bone in your skeleton. The muscles are the same – the more you exercise and use them, the more they grow and the stronger they get.

RECENT DEVELOPMENTS building muscles

Scientists have found that there are two different types of muscle fibres. "Slow twitch" muscle fibres contract relatively slowly and can work for long periods of time without tiring. They use oxygen and food very efficiently. "Fast twitch" muscle fibres contract very rapidly but tire more easily. Most people have about 50 percent of each type in their muscles. However scientists have found that Olympic athletes seem to have different arrangements of fibres that help them do well in different sports. Olympic sprinters have about 80 percent fast twitch fibres for bursts of speed, while marathon runners have up to 80 percent slow twitch fibres, which give them the stamina they need.

BROKEN BONES

When our skeletal system works properly, our joints move smoothly through a range of positions. But when things go wrong, the situation is painfully different.

It hurts when a bone breaks, and the broken bone doesn't work very well! Fortunately bone usually heals easily, although if it is a bone in a leg or an arm, it may have to be kept still in a plaster cast to allow it to mend in a straight line.

JOINT PROBLEMS

If a joint is suddenly wrenched in the wrong direction, it might become sprained. The ligaments that hold the bones together are torn, so they swell up and the whole joint becomes very painful. Treatment can range from a few days rest to a plaster cast to help it heal, depending on how bad the sprain is.

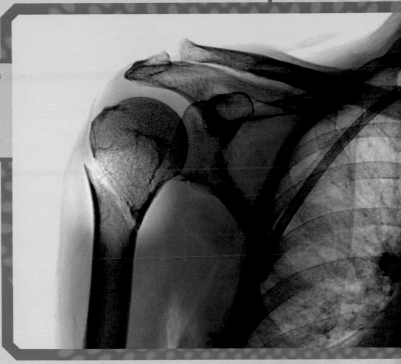

This X-ray image shows a broken humerus in the upper arm. It looks very dramatic, but soon the bone will be almost as good as new!

If a bone in a joint is actually forced out of its position, the joint is dislocated. This is incredibly painful, but can often be put right very quickly by putting the bone back into its proper place. If not, a **general anaesthetic** may be needed.

Osteoarthritis is a joint problem which often affects older people after a lifetime of wear and tear. The cartilage in the joints breaks down, so the bony surfaces rub together and movement becomes extremely painful. There is no cure, but an operation can replace some joints (such as the hip and knee) with an artificial one. These new joints are pain free and give their owners a new lease of active life!

Did you know..?

Bone is a living and everchanging tissue. Special cells, known as osteoblasts and osteoclasts, are constantly building up new bone where you need a little extra strength. They also break down bone where it isn't needed any more.

From puberty to parenthood

When we are small children, life is simple. But as adults we have to take on more responsibilities, and one of these may well be having children of our own. Reproduction, or producing more of the same kind, is something which is very important in the living world. Without it, living organisms would die out. In fact, for many animals and plants, reproduction is the only reason for their existence, and once they have reproduced successfully, they die. People, on the other hand, live for a very long time, so reproduction is only part of their life cycle.

BABIES — BOY OR GIRL?

When human beings are born they are very small and completely dependent on their parents to look after them. But even at this very early stage of life, the **reproductive organs** are there. Inside the body of a baby boy or girl are the cells that will later be involved in reproduction.

There appears to be very little difference between baby boys and baby girls, which is why people often dress boys in blue and girls in pink!

THE GROWING YEARS

As babies develop, they do a lot of growing. They get steadily bigger and as their bodies and their brains grow and mature, they can do more and more. Throughout childhood the bodies of boys and girls stay similar in shape and size. However, there are important differences. For people to reproduce successfully, genetic information has to be passed on from the parents to their offspring. Two specialized sex cells, one from each parent, have to join together to form a completely new individual. As boys and girls move towards their teenage years, their bodies begin to mature and change quite quickly so that they become capable of producing a baby.

SIGNALS OF CHANGE

The changes of **puberty**, which lead to sexual maturity, are very different in boys and girls, yet they are controlled in the same way. The triggers that tip you into puberty are complicated. It depends partly on the information you have inherited from your parents, but it is also affected by your body weight and how your own **hormones** work. All of the dramatic changes of growing up are controlled by these special chemical messengers made in your brain and by your sex organs – the **ovaries** and **testes**.

Did you know..?

By the time a baby girl is born, her tiny ovaries contain all the eggs she will ever have – several hundred thousand of them in each ovary! But only about 450 of the eggs will ever be released, and a tiny fraction of those will become babies.

GROWING UP FOR BOYS...

The time that boys start to grow up and enter puberty usually begins between the ages of 11 and 16. For some boys, all the changes happen quite quickly while for others the changes take years to complete. Everyone is different, and changes happen at different times. The changes start when hormones (chemical messages) from the brain make the testes start to produce the male sex hormone testosterone. The rising level of testosterone triggers the physical changes that occur during puberty.

BODY CHANGES

One of the first changes to occur during puberty is the **adolescent** growth spurt, when boys grow taller. This is when most boys catch up with and overtake their female classmates in height. **Pubic hair** and body hair also begin to appear, starting with the groin and the armpits. Some men are naturally smooth, but others are hairy and body hair may continue to develop for some years. Facial hair – a beard and moustache – begins to grow as well. It can take several years from the first growth of facial hair before a boy needs to shave every day, or a strong beard growth can develop very quickly.

When the Adam's apple (larynx) in a boy's throat gets larger, his voice becomes deeper. This is known as the voice breaking. This can happen gradually or very suddenly – in some cases a boy's voice changes overnight!

Another major change takes place in the sex organs (the **penis** and the testes). These grow larger and their skin darkens. The testes start to make **sperm**, the special cells that join with a female egg cell to make a baby. The sperm are carried in a special fluid called **semen**, which is full of nutrients for the sperm.

The male body shape also changes slowly during puberty. The bones and muscles grow and the shoulders and chest become broader than the hips. Boys also tend to gain muscle during puberty even if they don't do a lot of exercise.

HEAD CHANGES

It isn't just the body that changes during puberty – the brain changes too. Adolescents become more independent, more questioning and start to look out more and more beyond their family to their friends. At the same time adolescents can feel very young and insecure, or confused, or angry for no real reason. It is all part of growing up and changing hormone levels are at least partly to blame.

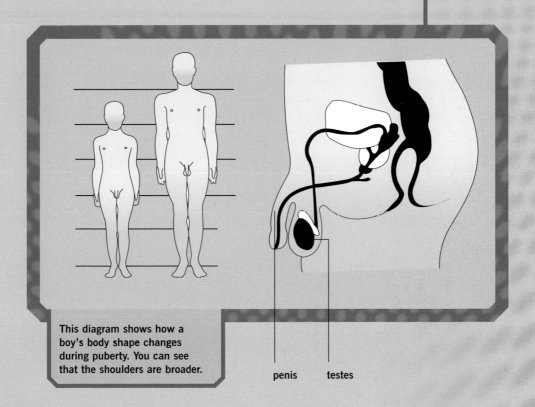

This diagram shows how a boy's body shape changes during puberty. You can see that the shoulders are broader.

penis   testes

Did you know..?

In normal healthy semen there are hundreds of millions of sperm. In fact the lowest number of sperm counted as normal is 20 million sperm per 1 cm^3 (0.06 in^3) of semen.

... AND GROWING UP FOR GIRLS

Girls often go into puberty slightly earlier than boys, with the changes beginning when they are between 10 and 15 years old. Just like boys, the timing and speed of puberty in girls varies greatly from one person to another. Although it is different for everyone – and everyone ends up a slightly different shape and size – the basic changes that take place are the same. Hormones from the brain make the ovaries start to produce the female sex hormone, oestrogen. As the levels of oestrogen rise and the body responds, all kinds of changes are triggered.

BODY CHANGES

Girls have a growth spurt during puberty, which takes them close to their adult height. The breasts develop, becoming larger so they are ready to produce milk for a baby when it is needed. The bones and muscles grow, particularly the **pelvis**. Girls develop fat around their hips, thighs and bottoms, which produces a more curvy, "feminine" body shape. And just like boys, pubic hair appears.

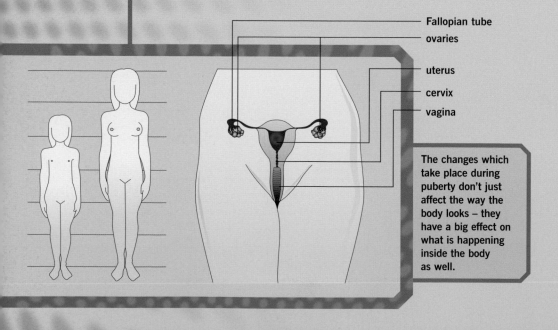

Fallopian tube
ovaries
uterus
cervix
vagina

The changes which take place during puberty don't just affect the way the body looks – they have a big effect on what is happening inside the body as well.

THE MENSTRUAL CYCLE

In some ways the most dramatic change to take place in a girl's body during puberty is the start of the **menstrual cycle**. Each month inside the ovaries, an **ovum** (or egg) starts to mature. When it is ready, the ovum is released from the ovary to travel along the **Fallopian tube** to the **uterus** (or womb). This is known as **ovulation**.

Each month the uterus prepares itself for pregnancy by growing a thick, blood-rich lining ready for an **embryo** to start developing. If no fertilized egg arrives in the uterus, the lining is shed and this blood loss is known as the monthly period. Although they are called "monthly", a girl's periods are not usually very regular when they first start – and for many women they never fall into a regular four-weekly pattern.

RECENT DEVELOPMENTS sex hormones and pain

Jon-Kar Zubieta, MD, PhD, and his colleagues at the University of Michigan in the USA have been studying the effect of sex hormones on how pain is felt in the brain. Not only have they found differences in the way men and women respond to pain, but also that women cope with pain much better when the levels of sex hormones are high in their monthly cycles than when they are low!

Did you know..?

The average girl loses 50 cm^3 (3 in^3) of blood at each menstrual period, and will have around 450 periods during her lifetime. This is approximately 22,500 cm^3 (1,370 in^3) of blood lost!

SO WHERE DO BABIES COME FROM?

Once the body has gone through puberty, it is prepared for sexual reproduction. However, most people wait until adulthood before having a child. The average age of first-time mothers in the United States is now around 25 years old – more than 10 years after puberty.

When a sperm penetrates an ovum, and the two sets of genes fuse, a unique human cell has been formed. If all goes well, in nine months time this will be a baby.

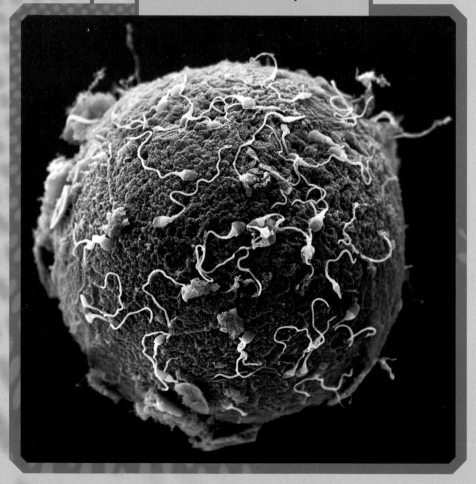

WHEN A SPERM MEETS AN EGG

The sperm are made inside the man's testes. The ova (eggs) mature inside the woman's ovaries. For human reproduction to be successful, a sperm and an ovum must join together, but to do that, they must meet. The sperm get inside the body of a woman during **sexual intercourse**.

Firstly, the sperm move from the testes, through the urethra in the penis and are then released inside the vagina in a process called ejaculation. Once inside the woman's body, the sperm then move through the uterus and up into the Fallopian tubes where the sperm will meet an ovum if the woman is at the right stage of her menstrual cycle. Of all the millions of sperm that set out on the journey, only a few hundred actually reach the ovum, and only one will actually fertilize the egg.

A CHANCE IN A MILLION

There are many difficulties to be overcome by the egg and the sperm before fertilization can take place. The ova cannot move along for themselves. Instead they are wafted along the Fallopian tube by the beating of lots of tiny hairs called cilia. What is more, the ovum only lives for about 24 hours after ovulation. The sperm have to travel through the **mucus** in the vagina and the cervix, then on through the uterus and into the Fallopian tube, containing the egg. Their journey is a bit like a person setting off from the United Kingdom and swimming to the United States, through treacle.

Did you know..?

Around 60–70 percent of fertilized ova never make it to be a baby – in fact around 50 percent of all fertilized eggs are lost before the woman misses a period. This is usually because there is something wrong with the embryo and it cannot develop normally.

PREGNANCY — A TIME TO GROW

Once fertilization has taken place, a woman becomes pregnant. The fertilized ovum starts to divide into a ball of cells as it travels on along the Fallopian tube, forming a tiny embryo. It enters the uterus and burrows into the blood-rich lining. Here it develops into a tiny but recognizable human being (known as a **foetus**). The umbilical cord connects the foetus to an organ called the placenta that provides blood and nourishment. Your tummy button is the spot where your umbilical cord attached you to your mother when you were growing inside her. After twelve weeks, the foetus has all its organs, although they are not fully formed or able to work on their own.

Food and oxygen pass from the mother to the foetus through the placenta, and the waste products made by the foetus are passed back to the mother. The placenta also acts as a defensive barrier, protecting the foetus from some of the diseases and harmful substances which may be in the mother's body. The foetus grows in a bag of fluid that cushions it from knocks and bumps and makes it easy to move around.

When a foetus has got this far through a pregnancy, almost everything it needs for independent life is in place. Now it just needs to grow and mature.

MAKING AN ENTRANCE

After around 40 weeks, the fully-grown foetus becomes too big for its mother's body. It is time for it to be born. The baby is pushed out of the uterus during **labour**. The muscles of the uterus contract strongly to open up the cervix (the neck of the uterus) and squeeze the baby out – usually head first – through the vagina. The time this takes can vary from under an hour to over 24 hours, and labour is usually hard and often painful work for the mother. Once the baby is born, the umbilical cord is cut, the placenta comes away and the baby has to breathe and start doing many things for itself.

RECENT DEVELOPMENTS
operating in the womb!

Identical twins often share a placenta and sometimes things go wrong. One twin can get too much blood while the other gets too little. Usually when this happens both twins die. But Dr Ruben Quintero in Florida has developed special keyhole surgery that he can carry out on affected identical twins in the uterus long before they are born. In about 50 percent of cases, both twins are saved, and in 84 percent of cases, at least one twin is saved. Now other doctors are trying his technique.

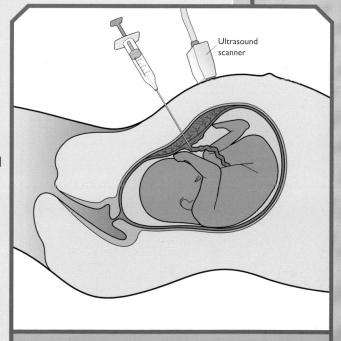

Ultrasound scanner

Twins aren't the only babies to benefit from surgery before birth. Blood transfusions into the umbilical cord like this one have saved the lives of thousands of babies whose blood is not compatible with their mother's.

PROBLEMS WITH FERTILITY

Many women get pregnant very easily when they decide to start a family. But up to one couple in six cannot conceive when they want to. There are lots of different reasons for this. The man may not make enough sperm, or any sperm at all, or his sperm may be faulty. The woman may not have any eggs, or she may have eggs but not release them each month, or the Fallopian tubes carrying eggs to the uterus may be blocked. For centuries there was no hope for infertile couples unless they were lucky enough to adopt a child. But infertility treatments have come a long way in recent years, and now a growing number of infertile couples are able to produce children of their own.

OVERCOMING INFERTILITY

Infertility can be treated in several ways. Women can be given artificial hormones that trigger their ovaries into releasing eggs. These fertility drugs increase the risk of having twins or triplets, but they have made it possible for thousands of couples to have children. However, the biggest breakthrough in infertility

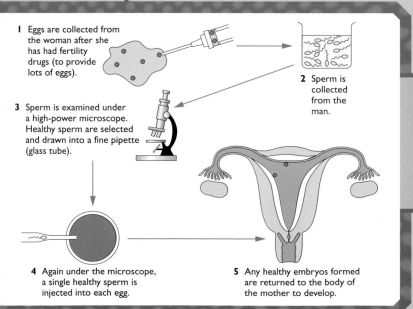

1 Eggs are collected from the woman after she has had fertility drugs (to provide lots of eggs).

2 Sperm is collected from the man.

3 Sperm is examined under a high-power microscope. Healthy sperm are selected and drawn into a fine pipette (glass tube).

4 Again under the microscope, a single healthy sperm is injected into each egg.

5 Any healthy embryos formed are returned to the body of the mother to develop.

A diagram like this makes IVF look simple. In fact it is a very complicated process that costs a lot of money. Most infertile couples who try IVF still do not end up with a baby.

treatment was the development of *in vitro* **fertilization** (also known as IVF or test-tube babies). This treatment combines the egg and sperm outside the mother's body, and returns the developing embryo to the womb so it can overcome the problem of damaged Fallopian tubes.

There are lots of other treatments which help overcome infertility, including injecting a single sperm into an egg, and freezing embryos, but they have all developed from the original groundbreaking work on IVF.

Cause of infertility	Treatment
No eggs in the ovaries (woman)	Egg donation for IVF
Not enough sex hormones (woman)	Fertility drugs
Fallopian tubes twisted, scarred, or blocked (woman)	Sometimes the tubes can be repaired. More often, women seek IVF treatment
No sperm (man)	Sperm donation (when donor sperm is used to fertilize a woman's eggs) or IVF treatment
Not enough sperm (man)	Stop smoking and drinking to raise sperm count, or sperm donation
Inactive sperm (man)	Injecting a single sperm into an egg, followed by IVF, or sperm donation

SCIENCE PIONEERS Edwards and Steptoe

In 1969, Dr Robert Edwards, of Cambridge University, fertilized a human egg outside the human body for the first time. In 1971 he started working with Mr Patrick Steptoe, a UK gynaecologist who pioneered a method of collecting ripe eggs from women. In 1978 they were responsible for the first test-tube baby ever born.

Louise Brown was conceived in a glass petri dish in 1977 and born in 1978. She brought fame to scientists Edwards and Steptoe – and hope to couples around the world!

Breathe in, breathe out

The first breath a baby takes when it is born shows that a new independent life has started. One of the first things we check for in an emergency is whether an injured person is breathing. When you stop breathing, you die. So why is breathing so important?

Every cell in your body needs energy to grow and carry out all the jobs it has to do to keep you alive. This means every cell needs oxygen, because oxygen is used to get the energy from your food during the process of **cellular respiration**. As the food is broken down in respiration, a poisonous waste product (carbon dioxide) is formed. Breathing is the answer to both of these problems – it brings oxygen into your body, and gets rid of waste carbon dioxide.

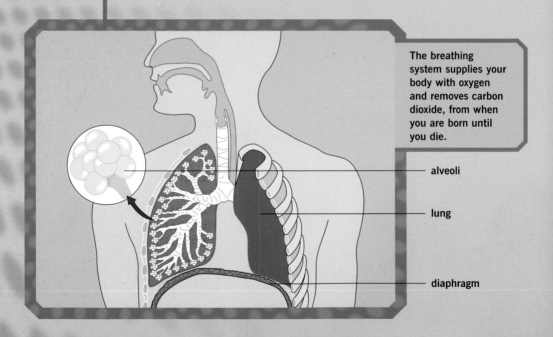

The breathing system supplies your body with oxygen and removes carbon dioxide, from when you are born until you die.

alveoli

lung

diaphragm

THE BREATHING SYSTEM

You breathe all the time, awake or asleep. Your chest rises and falls as air is moved into and out of your **respiratory system**. The air enters through your nose and mouth when you breathe in, and travels down a series of tubes until it reaches the lungs. Your lungs look like pink sponges, hidden and protected by the ribcage. It is in the lungs that oxygen is taken from the air to travel around the body to the cells. This is also where waste carbon dioxide is removed from the body as you breathe out.

The lungs themselves are made up of tiny air sacs called **alveoli**. These alveoli give the lungs an enormous surface area – if all the alveoli in your lungs were spread out, they would cover an entire tennis court! The alveoli have very thin walls with moist surfaces and lots of **blood vessels** in close contact with them. All of this makes it easy for oxygen to move from the air in the alveoli into the blood, and for carbon dioxide to move from the blood into the air. This process is known as **gas exchange**.

THE AIR WE BREATHE

Our bodies need to breathe in oxygen and to get rid of carbon dioxide, but the air we breathe in is not pure oxygen. Air is mainly nitrogen, a gas that has no effect on us at all!

Atmospheric gas	Air breathed in	Air breathed out
nitrogen	About 80%	About 80%
oxygen	20%	16%
carbon dioxide	0.04%	4%

Did you know..?

The volume of air an adult normally breathes in and out of their lungs is about 0.5 litres (2.4 pints) – but the biggest volume most people are capable of taking in is over 4 litres (9 pints)!

GETTING AIR INTO THE LUNGS

Air must be brought into your lungs and moved out again all the time, no matter what you are doing. This is done by movements of your ribcage (which you can see and feel) and by movements of your **diaphragm** (which you can't). As you breathe in, muscles contract to pull your ribs upwards and outwards. At the same time, your diaphragm contracts and flattens. The space inside your chest gets bigger and air moves into your lungs. As you breathe out, the muscles between your ribs and your diaphragm then relax, the space inside your chest gets smaller again, and the air is squeezed out of your lungs.

SCIENCE PIONEERS helping premature babies

In the 1950s and 60s, around 10,000–15,000 premature babies died each year in the US alone because they couldn't breathe properly when they were born. Dr Richard Pattle in the UK and Dr John Clements in the US discovered why. A special chemical called lung surfactant coats the surface of your alveoli, making it easy for air to move into and out of your lungs. When babies are born too early, their lungs don't make this chemical.

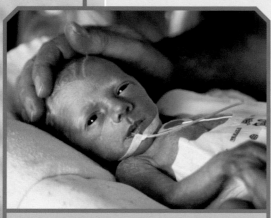

The work of scientists such as Tetsuro Fujiwara, John Clements, and many others like them has saved the lives of thousands of premature babies worldwide.

In 1980, Dr Tetsuro Fujiwara of Japan tried using surfactant taken from cow's lungs to help premature babies breathe – and it worked! Dr Mary Ellen Avery, a top US doctor, ran trials of the new chemical in the US, while John Clements worked on the first synthetic surfactant which could be made and used anywhere.

CONTROLLING YOUR BREATHING

Most of the time you breathe without thinking about it. This automatic control is very important. It would be hard to walk around, concentrate in school, or play sport if you had to think about breathing all the time, and no-one would dare fall asleep. But there are times when we need to take control of our own breathing – blowing up a balloon or swimming are just two examples. So our brains can override our automatic system and allow us to take deep, conscious breaths.

Without conscious control of our breathing, we could never play musical instruments like this.

LIVING AT HIGH ALTITUDE

People born at high altitude, where there is less oxygen in the air, have bodies specially adapted to get enough oxygen in and carbon dioxide out without breathing fast all the time. They have bigger lungs, with far more alveoli, than people born at sea level.

Did you know..?

A newborn baby has about 30 million alveoli. Before the time you are 10 years old this has increased to a final number of about 300 million!

LUNGS FOR LIFE?

Healthy lungs are important for the well-being of your whole body. When you exercise, your muscles work hard and so need more oxygen. You breathe harder and faster to supply that need. An average man will use around 200 cm^3 (12 in^3) of oxygen per minute when he is doing nothing, but around 4,000 cm^3 (244 in^3) if he is working flat out. If you exercise regularly, your body adjusts and becomes fitter, with bigger lungs and a better blood supply to them, so more oxygen reaches the body cells with every breath you take.

BREATHING PROBLEMS

Not everyone is lucky enough to have a respiratory system that works at its best. Asthma is becoming more and more common. It causes shortness of breath, wheezing, and coughing. At its worst, it is life-threatening. During an asthma attack, the muscles in the walls of the breathing tubes contract, the linings swell up, and extra mucus is made. All of these things make the tubes narrower, so it is difficult to move air in and out. All sorts of things can trigger an asthma attack, from cold air to animal hairs, and from exercise to house dust mites and cigarette smoke. Fortunately modern medicines make it possible for most asthma sufferers to lead active, healthy lives.

Inhalers allow medicines to be delivered straight to where they are needed. Most people affected by asthma will have an inhaler which opens up the tubes in the lungs, making breathing much easier when they are having an attack.

SMOKING AND THE LUNGS

One of the biggest enemies of healthy lungs is smoking. Cigarette smoke stops the system that clears dirt, mucus, and bacteria out of your lungs from working. Smokers are more likely to get infections from 'flu to **bronchitis** and **pneumonia**. Even more seriously, black, sticky tar and other nasty chemicals are left in the lungs, leaving smokers at risk of diseases like emphysema, where the alveoli break down to leave big air spaces. This means gas exchange in the lungs is much less efficient and sufferers are always short of breath. What's more, cigarette smoke contains many chemicals that cause cancer. As a result, smokers are more likely to die from lung cancer and throat cancer than any other group of people.

CLASSIC EXPERIMENT
cracking the smoking/cancer link

In 1951, Dr Richard Doll and Professor Austin Hill in the UK interviewed 1357 men with lung cancer, and found that 99.5 percent of them were smokers. This looked like strong evidence for a link. They set up a 10-year study, and in 1964 published results showing a dramatic fall in lung cancer cases in people who had stopped smoking compared to those who still smoked. Their work was vitally important in getting both doctors and the general public to understand the massive health risks posed by smoking.

Did you know..?

Around 87 percent of deaths from lung cancer, and 1 in 5 of all deaths in the US, are directly related to smoking – yet around 46 million Americans still smoke.

Transport to the cells

You are made up of billions of cells, and most of them are a long way from a source of food and oxygen. The body's circulatory system supplies all your cells with what they need for cellular respiration, and removes the waste products they produce. In this "transport" system your blood is moved around a system of pipes (the blood vessels) by a pump (the heart).

SCIENCE PIONEERS understanding circulation

The credit for working out how the heart and circulation work is usually given to the Englishman William Harvey in 1628. But Harvey was not the first person to explain how the heart works. In fact he was one of many, and he wasn't even the first one to get it right.

Doctors in China as long ago as the second century BC understood that the heart was a pump which moved the blood around the body. In fact, Chinese doctors used a model made of bellows and bamboo tubes to show their pupils how the heart and blood circulation worked.

William Harvey was an English doctor who studied his patients and living animals as well as dead bodies to find out what was going on in the heart and circulation. Harvey published his ideas about the heart in 1628, and everything we know today about the circulatory system is still based on his work.

Galen was a Greek doctor in the second century AD. He spent his whole life studying how the human body works – and still got it wrong! Galen believed that the heart sucked blood from the veins and that blood flowed from one side of the heart to the other through tiny holes. Unfortunately his wrong ideas were accepted for over a thousand years!

Ala'El-Deen Ibn-El-Nafis was a senior doctor in Cairo in 1284 AD. He worked out the correct structure of the heart and the way the blood flowed through it, but no-one else really understood his ideas. Around 300 years later, some of Ibn-El-Nafis' work was translated into Latin so it could be read in Europe. And a short time later, European scientists and doctors began to make the same discoveries! Was this a coincidence?

Michael Servetus (1511–1553) was a Spanish doctor who was trained in Paris. He wrote some very good descriptions of how the heart works, but unfortunately both Catholic and Protestant leaders did not approve of his religious writings. After Servetus escaped from capture by Catholics in Lyon, France, they burnt his effigy (a model of him). A few months later he was arrested, tried and executed in Geneva, Switzerland, by the Protestants.

Michael Servetus must be one of the only doctors ever to be put to death twice – and not even for his scientific ideas!

WHAT IS BLOOD?

If you cut yourself, the red liquid that oozes out is your blood. But blood isn't as simple as it looks. In fact, it isn't even really red. It is made up of a clear yellow liquid called plasma, that carries red blood cells, white blood cells, platelets, and lots of different chemicals. All the digested food molecules from the gut are carried around in the plasma to the cells where they are needed. Meanwhile, poisonous waste is moved where the body can get rid of it. Carbon dioxide is carried to the lungs, and **urea** from protein breakdown is moved to the kidneys.

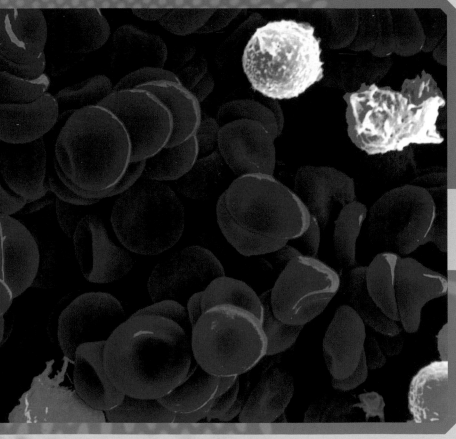

These different cells, mixed with plasma, make up your blood.

THE CELLS OF THE BLOOD

Blood gets its red colour from red blood cells. There are more of them than any other type of blood cell, and they have a very important job. Red blood cells carry oxygen from your lungs to all the cells, where it is used to get energy from your food. Inside the red blood cells is a special red pigment called haemoglobin. This picks up oxygen and carries it to where it is needed. When haemoglobin is carrying lots of oxygen, it is bright red. When it is not, it is purple-red. It is made using iron, which is why iron is such an important part of your diet.

White blood cells are much bigger than red blood cells, but there are far fewer of them in your blood. They help to defend you from disease by attacking bacteria and viruses that get into your body. They come in many different shapes and sizes.

Platelets are only small bits of cells, but they are a very important part of your blood. They help to make clots and repair your body when you cut yourself. Without platelets, you could bleed to death from a little cut.

CARRYING THE BLOOD

The blood is carried all over the body by a huge system of blood vessels. Arteries carry the blood away from your heart. Arteries have a pulse, so it is dangerous if they are damaged because the blood spurts out very quickly. Veins carry blood back towards the heart. They have valves to stop the blood flowing backwards round your body. Linking the arteries and the veins are the capillaries. These tiny blood vessels have very thin walls so that food and oxygen can pass easily out of the blood into the cells, and the waste products from the cells can pass easily into the blood.

Did you know..?

Each adult has approximately 5 litres (10.6 pints) of blood containing about 15 billion red blood cells that travel around their body in around 80,000 kilometres (50,000 miles) of blood vessels!

HOW YOUR HEART WORKS

Your heart is a muscle that beats about 70 times a minute all through your life, pumping blood around your body.

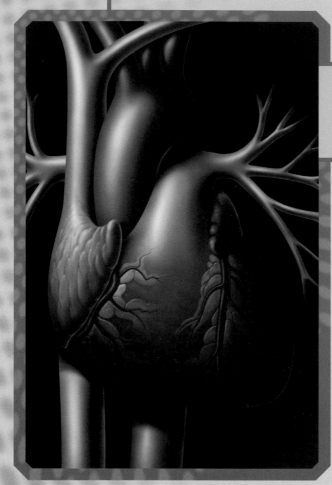

The human heart is only about the size of your closed fist but it pumps about 5 litres (10.6 pints) of blood around your body each minute, even when you aren't doing anything energetic!

WHEN THE HEART WORKS

The human heart is a double pump, with both sides filling and emptying together. The right side of the heart collects blood from the body and pumps it to the lungs to pick up some more oxygen and get rid of carbon dioxide. The left side of the heart collects this oxygenated blood from the lungs and pumps it all around the body.

Inside the heart, there are lots of different valves which let blood flow through in one direction, but close so the blood cannot flow backwards. The noise of the heartbeat heard through a stethoscope (a special instrument used by doctors) is the sound of these valves working as blood surges through the heart.

WHEN THINGS GO WRONG

Sometimes things can go wrong with the heart. Problems can vary from simple changes in the heart's rhythm (the beat going too fast or too slowly), through to heart attacks which can damage your heart so badly that it stops for good. Heart disease is a huge problem; in the US more people die from heart attacks than from any other single cause.

So what causes heart disease? There is no simple answer because the health of your heart is affected by so many things. However, these factors are important:

- what you inherit from your parents: a tendency to get heart disease often runs in families
- smoking: smokers are 70 percent more likely to die from heart disease than non-smokers
- food: a fat-rich diet, particularly animal fats, seems to increase the chances of heart disease by affecting the level and balance of **cholesterol** in your blood
- fitness level: regular exercise gives you fit and efficient heart muscles.

SCIENCE PIONEERS

In 1967, Dr Christiaan Barnard did the unthinkable when he removed the diseased heart from a very sick man and replaced it with a healthy heart from a 23-year-old woman who had died in a road accident. Although the patient only lived for 18 days, this pioneering first heart transplant paved the way for many other people. In the year 2000–2001, 160 heart transplants and 34 heart and lung transplants were carried out in the UK, while over 2000 heart transplants take place yearly in the US.

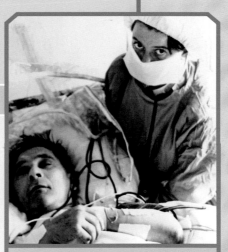

It was the courage of patients like Louis Washkansky, the first patient to receive a heart transplant – and the families who allowed organs from their loved ones to be used – that made the ground-breaking work of Dr Barnard possible.

FINDING OUT MORE ABOUT HEART ATTACKS

The heart is a very busy muscle, so it needs a good supply of blood to bring food and oxygen to the contracting muscle cells. Blood is brought to the heart muscle in blood vessels called the coronary arteries. If these become blocked, the blood supply to part of the heart muscle is cut off. That part of the heart muscle is starved of oxygen and dies. If too much of your heart muscle is affected, this can cause death.

Doctors and scientists have been working hard for years to try and find ways of preventing heart disease but there is no single answer. Scientists are still working on the reasons why it occurs in the first place.

When the arteries supplying your heart muscle with blood get clogged up with fatty substances like this, they get narrower and narrower until they are blocked completely – and this can cause a fatal heart attack.

RECENT DEVELOPMENT
heart health and the vitamin link

When Kilmer McCully was a young researcher at Harvard University in the US thirty years ago, he discovered a link between an amino acid called homocysteine and changes in the blood supply to the heart that can lead to heart attacks. High levels of homocysteine are linked to low levels of B vitamins in the diet. B vitamins are often missing in processed foods. Changing your diet or taking a supplement of B vitamins helps your body remove the homocysteine and prevents the damage to the heart.

Unfortunately McCully did his research at the same time as many top scientists were supporting the link between fats and heart attacks. Time and money had been spent developing anti-cholesterol drugs and low-fat foods. McCully lost his funding at Harvard and his ideas were ignored.

Thirty years later, and in spite of much treatment for raised blood cholesterol, deaths from heart disease are still high and Kilmer McCully's work is finally being taken seriously. Major trials on B vitamins and heart disease are taking place around the world. It seems increasingly likely that he has found one of the pieces in the jigsaw that explains heart disease.

SCIENCE PIONEERS   Hans Krebs and the mystery of cellular respiration

The circulatory system is very important because it delivers the oxygen needed in the incredibly complicated cycle of reactions known as cellular respiration which take place in the **mitochondria** of every cell. It also removes the carbon dioxide made as a waste product. The complicated biochemistry of cellular respiration was discovered by Sir Hans Krebs and his team in the UK during the 1930s. He worked out all the enzyme-controlled reactions which produce the energy needed for life.

Healthy bodies, healthy minds

Illness has been around as long as people have. Human remains from thousands of years ago show signs of diseases. But for most of that time people have not understood what causes diseases or how they can be cured.

Five thousand years ago, the Ancient Egyptians thought illness was caused when people upset one of their many gods. Treatments always included sacrifices to the gods, and substances like mouldy bread, poppy seeds, crocodile dung, and honey. Some of these cures may have actually worked because they contained chemicals that we now know do help to fight disease.

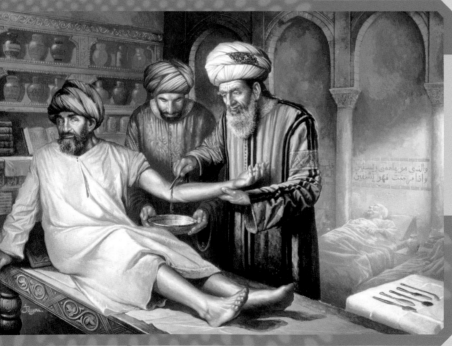

For centuries, no-one understood the causes of disease and the treatments owed more to folklore and magic than science. The "cure" was often as bad as the illness.

In ancient Greece, Hippocrates was convinced that disease had nothing to do with the gods. He thought that a healthy body contained a balance of four liquids – blood, black bile, yellow bile, and phlegm. If the fluids were out of balance, you were ill and needed treatment to restore the balance. His ideas were still accepted in 17th century Britain!

It was not until the 19th century that people began to have a scientific understanding of disease. Since then, helped by the development of more and more powerful microscopes, people have discovered bacteria and viruses. These **micro-organisms** are too small to be seen by the naked eye, but they cause all sorts of **infectious** diseases from tonsillitis to tuberculosis. Understanding the cause of diseases makes us better at treating them, and at preventing them from spreading in the first place.

SCIENCE PIONEERS understanding infectious disease

Many scientists have helped to unravel the mysteries of infectious diseases. These are just a few of them:

- Louis Pasteur showed that micro-organisms (germs) cause disease in people and made some of the first **vaccines** against infectious diseases in the 19th century
- Also in the nineteenth century, Dr John Snow realized that disease could be spread through water contaminated with micro-organisms. He stopped an outbreak of **cholera** in London
- In 1865 Joseph Lister introduced **antiseptics** – chemicals to kill bacteria – into hospitals
- Martinus Beijerinck discovered viruses in 1898 after he filtered bacteria out of a fluid, but found that the fluid still caused disease in plants
- Alexander Fleming, Howard Florey, and Ernst Chain discovered and developed penicillin, the first effective medicine against bacteria.

INFECTIOUS DISEASES

The most common diseases all over the world are the infectious diseases caused by micro-organisms such as bacteria and viruses. Bacteria are the smallest of all living things at around 0.005 millimetres long. They come in all sorts of shapes and sizes, and there are more bacteria than any other type of living organism. Fortunately, they don't all cause disease – the majority of bacteria do us no harm and some of them actually do us good. However the ones which cause disease can do terrible damage. Bacterial illnesses include pneumonia, tuberculosis, and the **plague**.

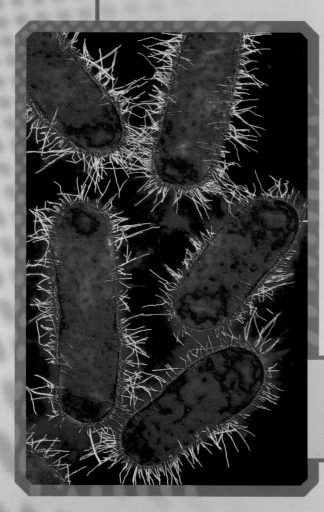

Viruses are incredibly small, at about 0.0001 millimetres long. They are made up of a protein coat containing a small amount of genetic material. They are not living things, but they act like living organisms in some ways because they can replicate (reproduce themselves) but only by taking over a living plant or animal cell. All viruses cause disease, including illnesses which range from the common cold to AIDS and polio.

It is estimated that in the United States, 73,000 people per year are affected by illness related to *Escherichia coli* bacteria (pictured here), 2,100 of whom are hospitalized, and 61 of whom die.

WHAT MAKES US ILL?

When bacteria and viruses get inside your body they reproduce very rapidly. Bacteria simply split in two – every twenty minutes if the conditions are right. Viruses, on the other hand, take over the cells of your body, use them to make lots of new viruses and then burst out destroying your cells on the way, ready to infect more cells. Both types of micro-organism damage your body tissues as they reproduce. It is this damage and the way your body responds to it (with fevers, aches, etc.) that makes you feel ill. Sometimes the micro-organisms produce toxins (poisons) as well, and the reaction of your body to the poison makes you unwell.

NATURAL DEFENCES

You are exposed to millions of micro-organisms every day but are not often ill thanks to the natural defences of your body. Your skin covers and protects most of your body tissues, and any micro-organisms that get in through your mouth are dropped straight into the acid of your stomach, which gets rid of most of them. The easiest way for micro-organisms to get into your body is through the respiratory system. However the hairs in your nose filter the air you breathe in, while many **microbes** get stuck in the sticky mucus in your lungs. This mucus is moved away from your lungs by special cells, covered with tiny beating hairs called cilia. When you cut yourself, your body copes by forming scabs which stop the bleeding. Scabs stop micro-organisms from getting in and also protect the new skin from further damage as it grows.

Did you know..?

In ideal conditions – plenty of food, warmth and oxygen and with no build up of waste products – a single bacterium will split in two every twenty minutes. It has been calculated that it would take only 44 hours, 2 minutes, and 42 seconds for the bacteria to weigh as much as the Earth. Fortunately bacteria don't usually find themselves in ideal conditions!

THE IMMUNE SYSTEM

In spite of all your defences, bacteria and viruses do sometimes get into your body. Fortunately you have a second line of defence – the immune system in your blood. You have an army of white blood cells (see page 43) that can swallow up and destroy invading micro-organisms. They make special chemicals called **antibodies**, which stick to a particular type of microbe and destroy it. Once your white blood cells can make the right antibody, they remember how to do it. You will never get that disease again – you are **immune**.

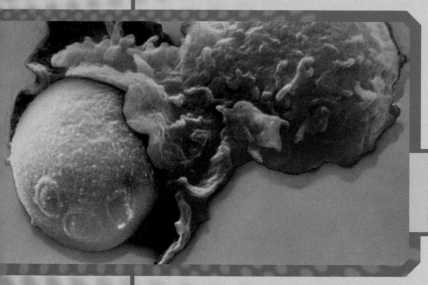

When microbes invade, sooner or later your immune system will wipe them out.

GIVING YOUR BODY A HELPING HAND

There are ways we can help our natural body defences deal with any **pathogens** we pick up. Destroying as many micro-organisms as possible before they get into your body is a good idea. Washing your hands after using the toilet and before handling food is a good way of doing this. You can also use antiseptics – chemicals which kill bacteria. Antiseptic creams and sprays help stop cuts and grazes from getting infected, and antiseptics are also widely used by hospitals, dentists, and vets. Unfortunately however, antiseptics have no effect on viruses.

CLASSIC EXPERIMENTS Semmelweis and experimental hand washing

In the late 19th century, when women began having babies in hospitals, childbed fever became more and more common. It was a dreadful infection that affected women shortly after giving birth, often killing them. No-one knew the cause.

Ignaz Semmelweis was a doctor working at the Vienna General Hospital with two maternity wards, one staffed by midwives and one by medical students. Semmelweis noticed that at least 12 percent of the women helped by the students died of childbed fever, three times as many as in the ward with midwives. He also noticed that the students often went straight from dissecting a dead body to delivering a baby without washing their hands. He decided they might be carrying the disease with them.

Then another doctor cut himself while working on a dead body, and died of symptoms identical to childbed fever. Now Semmelweis was sure that childbed fever was infectious and set up an experiment. He insisted all his students wash their hands in chlorinated lime before they even entered the maternity ward, and within six months the death rate had dropped to 1.27 percent. Sadly it was many years before his evidence was widely accepted. Even today, thousands of people die each year from infections they have picked up in hospitals.

As many as 1 in 3 new mothers used to die of child-bed fever. Modern levels of hygiene and antiseptics mean that in countries like the US and the UK, childbed fever is now extremely rare.

ANTIBIOTICS – A MEDICAL MIRACLE

In spite of all our precautions, we still get ill from time to time. Fortunately we have medicines called antibiotics that can kill bacteria faster than our immune system. They are especially important for serious diseases like tuberculosis and septicaemia, but they are also useful for conditions like tonsillitis. Antibiotics have saved millions of lives around the world, but they only work on diseases caused by bacteria. They have no effect on viruses.

RECENT DEVELOPMENTS crocodile blood and fish slime

In Australia, scientists have extracted a chemical they call crocodillin from the blood of wild crocodiles. It stops the crocodiles from getting infected if they lose a limb in a fight, even though they live in filthy water. Can it help make people better too?

In the UK, fish slime has been identified as another possible source of antibiotics. Not as exciting as crocodiles, perhaps, but trout are a lot easier to handle.

Penicillin, the original antibiotic, was found by accident when it was produced by a mould growing on a laboratory culture. Scientists today are looking a bit further out from the lab door in their quest for new antibiotics. Getting a blood sample from a crocodile is definitely more of a challenge than growing mould!

VACCINATION VICTORY

Some diseases can kill. Others, like measles and polio, can cause permanent damage. Antibiotics can be used to treat a disease when a person is already ill, but an even better idea is to prevent the person from getting ill in the first place. Immunization (or vaccination) does just that. Immunization introduces your body to these dangerous micro-organisms in a safe way. The vaccine contains dead or weakened bacteria or viruses. Your white blood cells can then learn to recognize these harmless pathogens and develop antibodies against them. So if you ever meet the real thing, your body will be able to make the right antibodies needed to destroy the micro-organisms before they can make you ill.

Most children in the developed world are now immunized against a whole range of diseases when they are very young. These include tetanus, polio, diphtheria, whooping cough, measles, mumps, rubella, and meningitis C. As a result, the death of a child from one of these once common infectious diseases has become rare.

SCIENCE PIONEERS Louis Pasteur

Louis Pasteur (1822–1895) was a successful scientist but he was broken-hearted when three of his five young children died of diphtheria. He was determined to do something about it. Pasteur convinced people that infectious diseases were caused by micro-organisms. He developed vaccines against several diseases, including anthrax and rabies. By the end of his life, Pasteur was close to finding a vaccine against diphtheria, the disease that killed his own little girls. His work still continues in Pasteur Institutes around the world.

THE CHOICE IS YOURS

You don't have much control over the infectious diseases you catch. But many people suffer from serious illnesses – and even die – as the result of choices they make about the way they live their lives.

The most common way in which people damage their bodies is by using drugs. A drug is a chemical that has a specific effect on your body. Common drinks such as tea, coffee, cola, and hot chocolate all contain the legal drug caffeine.

LEGAL DANGER

The two drugs which cause the most damage to health and the highest number of deaths each year are nicotine – found in cigarettes – and ethanol (alcohol) – found in alcoholic drinks. Like many drugs, nicotine and alcohol are addictive.

People get nicotine through smoking, and the hundreds of chemicals in cigarette smoke cause lots of health problems from lung cancer to heart disease.

Drinking alcohol in small amounts can be good for your heart. Your liver can break down small amounts of alcohol, but it can't cope with big amounts. It just gets damaged and eventually destroyed. Alcohol damages your brain tissue as well. It is very addictive.

Despite the health problems that they cause, both of these drugs are legal and in common use.

There are several ways that people try to give up smoking. Because nicotine is addictive, this is often very hard to do – but the health risks of continuing to smoke are strong motivation for many people.

ILLEGAL DRUGS

There are some drugs which are not legal because they are considered too dangerous, even in limited use. It is against the law to use them, yet some people still choose to put them into their bodies. Many of these drugs have a powerful effect on your mind, often producing vivid waking dreams called hallucinations. Some of these drugs are very addictive, and they can have terrible effects on your health, for several different reasons. They can cause many problems including:

- Sudden death: some illegal drugs can cause massive heart failure and death, even the first time you try them.
- Disease: AIDS and hepatitis are just two of many diseases which can be caught when addicts share the needles they use to inject drugs into their blood.
- Mental health problems: because drugs affect your mind, they can trigger serious mental health problems.
- Money problem: illegal drugs are expensive so addicts often turn to crime, and spend money on drugs instead of food. This can cause serious health problems and many addicts are badly undernourished.

Did you know..?

It is estimated that 23 million Americans are addicted to substances like alcohol, nicotine, and cocaine. It costs around US$400 billion each year to deal with all the health and social problems these addictions cause.

Tomorrow's research — today

All around the world, scientists and doctors are carrying out research into the way the systems of the human body work, and looking for new ways to improve our health and cure diseases.

ALCOHOLISM — WHAT NEXT?

Scientists have observed that alcoholism tends to run in families. Research carried out in the 1970s on twins showed clearly that there is an inherited link. Now scientists have broken down the whole of the human **genetic code**, they are hoping to pin down exactly which genes make a person prone to addiction. Other scientists are concentrating on special molecules in the brain that respond to alcohol. They are working on developing medicines that would block the alcohol and stop it having any effect. In the future, a combination of genetics, medicines, and counselling will enable many more addicts to be freed from their addiction, saving both money and lives.

ESCAPING INFERTILITY

Another busy area of research is in new fertility treatments. In 2004 a baby girl was born to Ouarda Touirat in Belgium. Ouarda had received treatment for cancer which damaged her ovaries, preventing her from being able to become pregnant. However, before the cancer treatment started, doctors removed a piece of healthy ovary tissue. Once Ouarda was free from cancer, they replaced the tissue near one of her ovaries. A few months later, she was pregnant! The freezing of human eggs and ovary tissue is an exciting way forward for fertility treatment.

THE STEM CELL DEBATE

In 1998, two teams of US scientists managed to culture human embryonic stem cells, which can form almost any tissue in the body. These cells might eventually be used to provide new body tissues. They could provide new nerve cells for people with spinal injuries, **Parkinson's** or **Alzheimer's**, or new heart muscle cells for people who have suffered a heart attack. They may even be able to provide whole new organs for people who need transplant surgery.

However, there is a debate within the medical community, the government, and the public about whether medical research should use stem cells. Some of the embryonic stem cells come from "spare" embryos in infertility treatment and others are from aborted embryos. There are many people, including many religious groups, who feel that using these cells is wrong. In the US, funding has been restricted so that it is very difficult for stem cell research to continue. In the UK, Australia, and most of Europe, research is allowed but there are strict guidelines to control the research.

In the future, scientists may be able to use stem cells from alternative sources, such as umbilical cords or from adults, instead of from embryos. Whatever happens, it is likely that stem cell research will join other medical breakthroughs in leading to better health for everyone.

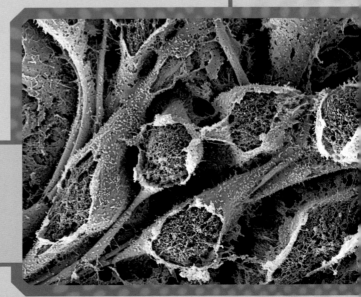

Stem cells are at the heart of one of the great scientific breakthroughs of the 21st century. The possibility of using stem cells to grow new tissues and even new organs is very exciting – but also, in some countries, very controversial.

Further resources

MORE BOOKS TO READ

Greenberg, Keith Elliot, *Stem Cells* (Blackbirch Press, 2003)

Parker, Steve, *Our Bodies* series (Hodder & Stoughton Childrens, 2003)

Stockley, Corinne, *The Usborne Illustrated Dictionary of Biology* (Usborne Publishing, 2005)

Nature Encyclopedia (Dorling Kindersley, 1998)

USING THE INTERNET

Explore the Internet to find out more about body systems and health. You can use a search engine, such as www.yahooligans.com or www.google.com, and type in keywords such as *digestive system*, *balanced diet*, *puberty*, *human respiration*, or *stem cells*.

These search tips will help you find useful websites more quickly:

- Know exactly what you want to find out about first.
- Use only a few important keywords in a search, putting the most relevant words first.
- Be precise. Only use names of people, places, or things.

Disclaimer

All the internet addresses (URLs) given in this book were valid at the time of going to press. However, due to the dynamic nature of the Internet, some addresses may have changed, or sites may have ceased to exist since publication. While the author and publishers regret any inconvenience this may cause readers, no responsibility for any such changes can be accepted by either the author or the publishers.

Glossary

adolescent   developing from a child into an adult

alveoli   microscopic air spaces in the lungs

Alzheimer's disease   form of dementia most likely to occur in older people

amino acid   building block of proteins

antibody   chemical made by the immune system which destroys pathogens

antiseptic   chemical that prevents the growth of disease-causing micro-organisms

anus   ring of muscle that closes the gut at the bottom end

bacteria   type of micro-organism that can be helpful, but that can also cause disease

blood vessel   tube that carries the blood around the body, such as arteries, veins, and capillaries

bronchitis   bacterial infection of the bronchi (tubes leading to the lungs)

cellular respiration   series of chemical reactions by which glucose is broken down using oxygen to produce energy for the cell. Carbon dioxide and water are the waste products.

cholera   infectious disease of the intestine

cholesterol   substance that is present naturally in the body, but can cause problems in large amounts

circulatory system   system made up of the heart and blood vessels that carries the blood around the body

diaphragm   muscle which helps us to breathe

digest   break down into smaller molecules

digestive system   system made up of a set of organs that digest food into the molecules needed by the body

embryo   baby at a very early stage of development inside the mother

enzyme   protein molecule that changes the rate of chemical reactions in living things without being affected itself in the process

Fallopian tube   part of the female reproductive system linking the ovary and the uterus

fat   very high energy food used as a source of energy and to store energy

fatty acid   building block of fats

foetus   unborn baby more than eight weeks into development

fungi   kingdom of organisms that do not move around and cannot photosynthesise

gas exchange   process of oxygen being taken into the body and carbon dioxide being given off

gene   unit of information in the DNA

general anaesthetic   chemical which causes the loss of sensation and consciousness before surgery

genetic code   information carried by the DNA

genetic information   information contained within the chromosomes located in the nucleus of each cell

glucose   simple sugar made by photosynthesis and used as a fuel in the cells of the body

glycerol   molecule formed when food is broken down. Glycerol can be converted to glucose by the liver and provide energy.

hormone   chemical message carried around the body in the blood

immune   protected against a disease

infectious   means something can be easily transmitted

laboratory culture   controlled growth, for example bacteria

labour   process by which a fully developed foetus is delivered from inside the mother

locust   insect that can form massive swarms which destroy all the plant life in an area

malnutrition   condition that comes from a lack of a balanced diet

membrane   very thin 'skin' which covers all cells and controls what moves into and out of the cell

menstrual cycle   part of the female reproductive cycle when the lining of the uterus is shed

microbe (see **micro-organisms**)

micro-organism   bacteria, viruses, and other minute organisms which can be seen only using a microscope

mitochondria   found in the cytoplasm of cells, where the reactions of respiration take place providing energy for the cell

molecule   group of atoms bonded together

mucus   substance produced by animals for lubrication or protection

muscle   tissue made up mainly of protein. It consists of long cells which shorten (contract) when stimulated.

organ   part inside the body which performs a particular job. Organs working together make systems within the body.

ovary   part of the female reproductive system that produces eggs

ovulation   point in the female reproductive cycle when an egg is released from the ovary

ovum   female sex cell in animals (also known as an egg)

Parkinson's disease   disease which affects the brain and causes increasing levels of shaking and disability

pathogen   disease-causing micro-organism

pelvis   part of the body at the base of the spine

penis   part of the male reproductive system

peristalsis   squeezing movements of the gut muscles which move food through the digestive system

placenta   part in the uterus which connects to the developing foetus to provide food

plague   contagious bacterial disease

pneumonia   bacterial infection of the lungs

protein   substance which living things need to grow new cells and replace old ones

puberty   stage of human development when the sexual organs and the body become adult

pubic hair   hair that begins to grow around the genital area during puberty

reproductive organ   organ of the body which is involved in reproduction

respiratory system   system made up of a set of organs that take oxygen into the body, and remove carbon dioxide

semen   fluid which contains the sperm

sexual intercourse   process by which sperm are passed from the man to the woman

specialize   adapt to carry out different functions

sperm   male sex cell in animals

tendon   strong tissue which joins muscles to bones and does not stretch at all

testes   part of the male reproductive system which makes sperm

tissue   specialized cells grouped together to carry out a function

urea   chemical produced by the breakdown of proteins in the body and excreted in the urine

uterus   part of the female reproductive system where a baby develops

vaccine   substance that causes the body to make more antibodies to protect against a particular disease

Index